Regional Traffic Signal Operations Programs: An Overview

October 2009

U.S. Department of Transportation
Federal Highway Administration
Office of Operations
1200 New Jersey Ave., SE
Washington, DC 20590

1. Report No. FHWA-HOP-09-007	2. Government Accession No.	3. Recipient's Catalog No.	
4. Title and Subtitle Regional Traffic Signal Operations Programs: An Overview		5. Report Date October 2009	
		6. Performing Organization Code	
7. Author(s) Principal Investigator: Peter Koonce Co-Authors: Kevin Lee and Tom Urbanik		8. Performing Organization Report No. Project 7372.02	
9. Performing Organization Name and Address Kittelson & Associates, Inc. 610 SW Alder Street, Suite 700 Portland, OR 97205 Subconsultants: Science Applications International Corporation (SAIC)		10. Work Unit No. (TRAIS)	
		11. Contract or Grant No. Contract No. DTFH61-06-D-00005, Task Order No. 4400149309	
12. Sponsoring Agency Name and Address U.S. Department of Transportation Federal Highway Administration Office of Operations 1200 New Jersey Ave., SE Washington, DC 20590		13. Type of Report and Period Covered Final Report March 2006 to October 2009	
		14. Sponsoring Agency Code HOP	
15. Supplementary Notes Eddie Curtis (Eddie.Curtis@fhwa.dot.gov) was the Technical Representative for the Federal Highway Administration. Additionally, Paul Olson of FHWA participated as support, providing comments throughout the project.			
16. Abstract This report provides an overview of practices related to developing and sustaining a Regional Traffic Signal Operations Program. The purpose for a Regional Traffic Signal Operations Program is to provide regional partners a formal framework to collectively manage the signal system performance for efficiency and consistency. A key benefit of a regional program is the development of projects that are of a magnitude that they can be included in a regional or state transportation improvement program (TIP). There are many benefits to the development of a regional traffic signal management and operations program. Agencies and users benefit from regional traffic signal operations programs as planners, engineers, and operators can provide an effective and efficient traffic signal system to the public and also provide higher levels of customer service without increasing costs. Additionally, by sustaining collaboration, regional operators can demonstrate to the public and elected officials that progress is being made on community goals, which then can be leveraged for future funding. Agencies and jurisdictions within a region that use a common framework for developing and establishing expectations, managing resources, and building relationships will result in more successful systems both individually and region-wide.			
17. Key Words Regional traffic signal management and operations, regional programs, signal timing, Regional operations, transportation system management and operations, planning for operations, operations strategies, regional concept for transportation operations, Policy Based Signal Control, Maintenance and Operation of Traffic Signals		18. Distribution Statement No Restrictions. This document is available to the public	
19. Security Classif. (of this report) Unclassified	20. Security Classif. (of this page) Unclassified	21. No. of Pages 53	22. Price N/A

Table of Contents

Section 1 - Introduction 1

Section 2 - What is a Regional Traffic Signal Operations Program? 7

Section 3 - Common Threads 25

Section 4 - Framework for a Regional Traffic Signal Operations Program 29

Section 5 - Maintaining a Sustainable Regional Traffic Signal Operations Program 37

Section 6 - Summary 53

Section 1
Introduction

Introduction

Overview of the Report

In response to the 2005 and 2007 Traffic Signal Report Card, the FHWA initiated Regional Traffic Signal Operations peer reviews to take an in depth look at the practices and processes that guide traffic signal management. The outcome of the peer reviews are a list of observations and recommendations that provide a path for improving the management, operation and maintenance of traffic signal systems. The peer reviews demonstrated that there are many common issues for communities across the country that can be addressed through regional cooperation. One of the most telling is that while there is an understanding that there is a significant need for more funding, there has not been a common voice by which to request additional attention. The initial intent of the Regional Traffic Signal Operations Program was to capture proven techniques, agreements, and opportunities that can be used by agencies to improve what they do. This document summarizes the Regional Traffic Signal Management Workshop hosted for FHWA where common threads were discussed. These common threads focused on the people, the processes, and the specific activities that have sustained successful programs across the country. The workshop provided a foundation for this report.

This report also documents other FHWA efforts that evaluate and characterize regional transportation management and operations and the importance of integrating them into the planning process. The overarching themes of collaboration and coordination discussed in these documents are essential to implementing and sustaining these broader

> "By combining efforts in a Regional Traffic Signal Operations Program, a common voice can be found."

regional transportation system operations programs. These regional transportation system operations efforts consist of activities ranging from highway incident management to transit mobility. There are many potential stakeholders beyond the normal participants, including emergency responders, public safety, transit agencies, economic interests, and freight operators. A common theme that binds all of these stakeholders is the mission of effectively and efficiently moving people and goods.

The findings from the assessments and workshop described herein outlines a framework for developing and sustaining a Regional Traffic Signal Operations Program. The elements described in this report are recommendations for use by managers and practitioners to improve the quality of traffic signal management on a regional level through an objective oriented process that engages stakeholders and leverages collaboration to achieve success.

Background

Traffic signals are an ubiquitous part of the transportation system that garner little attention when compared to the congestion and operational issues that surround freeways and highways in most medium and large metropolitan areas. Considering the significant impacts that traffic signals have on mobility, congestion, fuel consumption and climate change, a grade of "D" on the 2007 National Traffic Signal Report Card, indicates that there is significant room for improvement. An estimated 80 percent of all traffic signals in the United States are managed, operated and maintained at the local agency level. There are more than 2000 separate agencies responsible for traffic signal management and operation throughout the United States, with a significant percentage of these responsible for fewer than 50 traffic signals. Agencies that manage small numbers of traffic signals are very unlikely to have staff with a proficient level of technical expertise to effectively manage and operate traffic signals. This becomes even more problematic in an environment where little documentation and training resources exist to guide these activities. These smaller agencies also lack the political support needed for an effective program.

While decentralization of public services to the local agency level is normal in the United States it has had a negative effect on the management, operation, and maintenance of traffic signals. Without the expertise to adequately assess the needs of traffic signal systems, agencies typically focus primarily on maintenance and by default forgo many operational enhancements that could greatly improve the traffic signals. Proactive management and operation of traffic signals maximizes surface street capacity, which has the potential effect of offsetting the need to add capacity through costly construction activities.

An aging and retiring workforce is quickly depleting the expertise available to manage traffic signal systems, and the point of diminishing returns has been reached in terms of work force reduction and technology enhancements. A number of agencies have realized that a collaborative environment of regional traffic signal management, operation, and maintenance can leverage strengths and minimize organizational weaknesses. The people, processes, and procedures related to traffic signal management can be greatly enhanced by focusing priorities and establishing relevant objectives that benefit the region.

Regional Traffic Signal Operations Programs are not a new concept. Recently, a diversity of organizational structures, institutional arrangements, training programs, and collaborative resource sharing has resulted in better management, operation, and maintenance practices for traffic signal systems, which benefit the entire transportation network. Because traffic signals are typically located in areas with a significant number of users, their impact on mobility can be significant if not operated well.

In an era of shrinking budgets and limited resources, agencies have recognized the opportunity to make improvements by leveraging their strengths and the needs of partner agencies to increase customer service, enhance operations and maintenance activities, and improve the training and professional development of their staff. Integrating management and operations into the planning process allows a region to maximize the potential benefits of transportation investments. With respect to traffic signal systems, the regional portion of the program seeks to increase cooperation and communication among partner agencies to provide effective management, operations, and maintenance of traffic signals within a region.

Improving the operation of traffic signals involves much more than periodically updating signal timing. Active signal management is the outcome of supporting the people, process, and programs that manage, operate, and maintain the system. This requires establishing specific and measurable objectives. The case studies presented in this document highlight the common and critical elements that are essential to developing, implementing, and sustaining regional traffic signal operations programs.

> The Congestion Management Process is required to be developed and implemented as an integral part of the metropolitan planning process in Transportation Management Areas (TMAs) – urbanized areas with a population over 200,000.
>
> The eight steps involved in the Congestion Management Process include relevant activities that are directly related to traffic signal operations.

This document is one step in introducing and framing the concept of Regional Traffic Signal Operations Programs. Subsequent work is planned to develop additional resources and tools to facilitate the implementation of comprehensive regional programs. These will address multimodal, intermodal, and cross-jurisdictional systems, services, and projects designed to preserve capacity and improve security, safety, and reliability of the transportation system.

Definition of Traffic Signal Management

Traffic Signal Management is the planning, design, integration, maintenance, and proactive management of a traffic signal system in order to achieve policy based objectives to improve the efficiency, consistency, safety, and reliability of the traffic signal system. This includes the design and maintenance of timing parameters for the traffic conditions as well as the maintenance of the equipment. Traffic signal systems include a wide variety of subsystems, such as traffic signal displays, traffic signal controllers, detection systems, data-collection and archiving, surveillance and monitoring, and telecommunications.

By extending this across jurisdictional boundaries and cooperating amongst agencies, there can be effective collaboration to improve service quality by sharing experiences and planning to address future needs.

> "A Regional Traffic Signal Operations Program can be accomplished though interagency agreements where multiple agencies share responsibilities."

FHWA Traffic Signal Operations Reviews

The FHWA Resource Center Operations Technical Service Team has conducted Regional Traffic Signal Operations Reviews in thirteen regions throughout the United States, including Puerto Rico. These reviews have focused on benchmarking current traffic signal management practices and providing recommendations to advance traffic signal operations in the region. A primary recommendation that has emerged from all thirteen of the reviews is that funding for traffic signal management and operations is inadequate and could be improved through regional collaboration and management of the traffic signal system. In each region the MPO has taken a lead role in the development of a Regional Concept of Transportation Operations (RCTO) and provides a framework for regional partners to articulate objectives, performance measures, strategies, and agreements that improve traffic signal management and operations. By grouping traffic signal projects from multiple jurisdictions and demonstrating how collaboration enhances consistency, reduces redundancy, and improves the quality of traffic signal timing, a strong case is made for investing resources into a sustainable, Regional Traffic Signal Operations Program.

The RCTO framework allows operators and planners to discuss how improvements and a collaborative operation of the traffic signal system lead to regional benefits. These can be measured physically with reduced air pollution and delay fuel consumption as well as with institutional benefits that improve the productivity of the organizations involved in the day-to-day management of traffic signal systems. Safety benefits are also evident as a result of Regional Traffic Signal Operations Programs by promoting consistency and adding a level of accountability in how traffic signals are designed, operated, and maintained.

Model Regions for Regional Traffic Signal Operations Programs

- Denver, Colorado - Denver Regional Council of Governments (DRCOG)
- Kansas City, Missouri – Mid America Regional Council (MARC)
- Las Vegas, Nevada – Regional Transportation Commission of Southern Nevada (RTC)
- Puget Sound, Washington – Puget Sound Regional Council (PSRC)
- Reno, Nevada – Washoe County Regional Transportation Commission (RTC)
- Tucson, Arizona – Pima Association of Governments (PAG)
- Los Angeles, California – Los Angeles County Metropolitan Transportation Authority (METRO)

Consideration of regional traffic signal management and operations strategies in the metropolitan planning process has produced significant benefits to transportation system users and operating organizations within the region. Regional training programs, outreach, maintenance, management and operational benefits have resulted from Regional Traffic Signal Operations Programs. The programs are cost effective and low cost relative to infrastructure and capacity expansion projects that have long been a staple of the metropolitan planning process. A future of even more tightly constrained transportation budgets, fewer infrastructure projects, and a focus on retrofitting and maintaining current infrastructure is fertile ground for regional traffic signal management.

Purpose of a Regional Traffic Signal Operations Program

The purpose of a regional program is to provide partner agencies a formal framework to discuss issues, plan for improvements, and share experiences. By working across traditional jurisdictional boundaries, the agencies can provide higher levels of customer service through more objective oriented actions that consider the regional impacts of local activities. The concept of linking activities amongst agencies is consistent with the direction of Metropolitan Planning Organizations (MPOs), which lead to projects of a magnitude that they can be considered as a part of the Transportation Improvement Plan (TIP). Regional relationships also result in shared experiences that lead to improved practices and efficiencies in standard functions without increasing costs. Additionally, by sustaining collaboration, regional operators can demonstrate to the public and elected officials that progress is made toward meeting community goals, which can be leveraged for additional funding. Agencies and jurisdictions within a region that use a common framework for developing and establishing expectations, managing resources, and building relationships will result in more successful systems both individually and region-wide.

Section 2
What is a Regional Traffic Signal Operations Program?

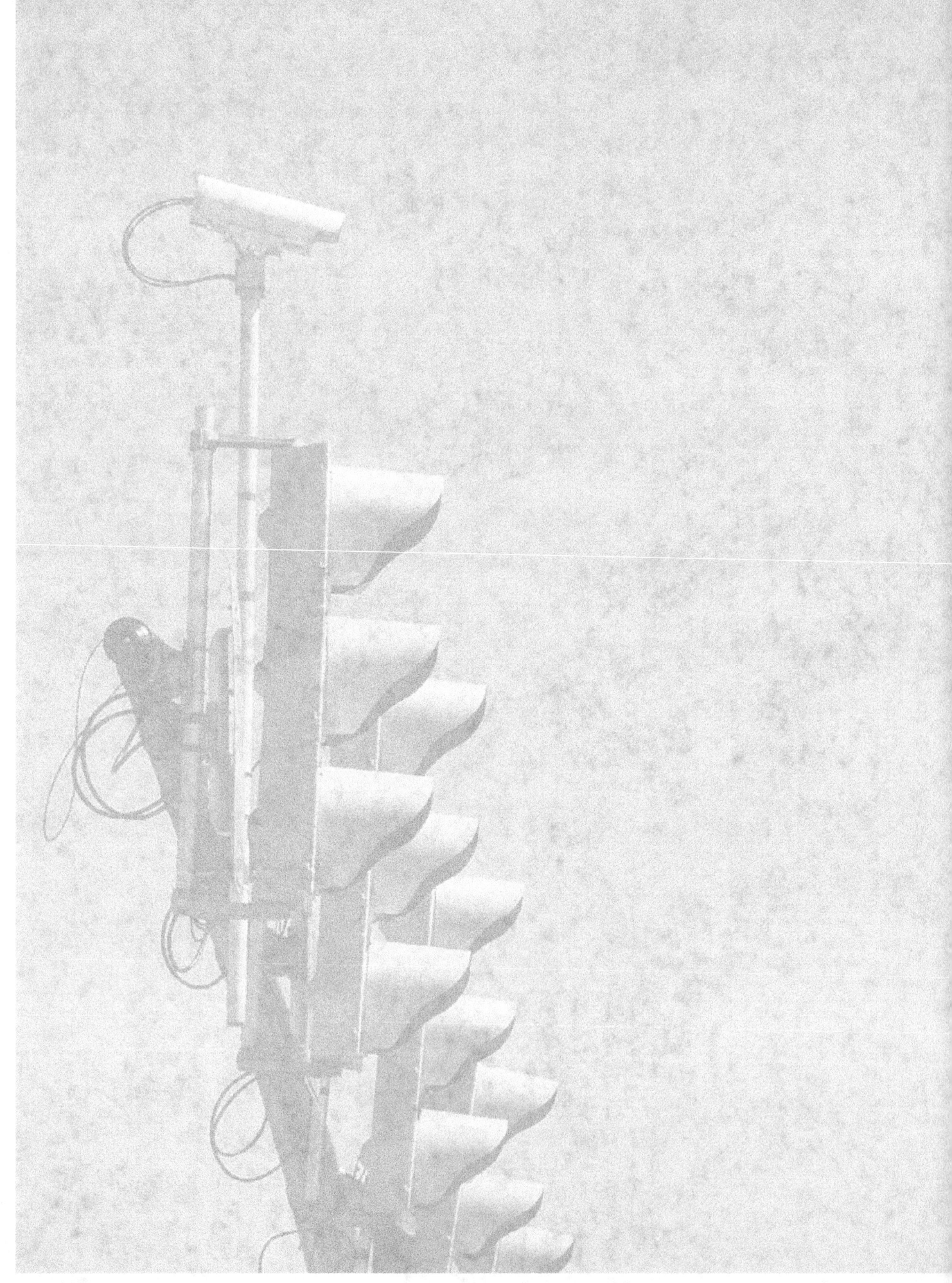

What is a Regional Traffic Signal Operations Program?

A Regional Traffic Signal Operations Program is a specific component of the broader program of Regional Transportation Operations Collaboration and Coordination[1], which makes the case to achieve safe, reliable, and secure transportation. This introductory document encourages and enables regional operations collaboration for transportation managers and public safety officials from states, counties, and cities within a metropolitan region. These managers and officials include traffic engineers, planners, transit operations, police, fire, and other emergency services, and port authority (e.g., air and water) managers. The primer can help these managers and officials understand what regional operations collaboration and coordination means, why it is important, and how to get started. In many cases, this document will also help those agencies currently engaged in some aspects of regional operations build on what they are already doing and expand it to include a focus on traffic signals.

Why a Regional Program?

Public agencies responsible for operation and management of traffic signals have limited resources and staff redundancy to guarantee continuity over time. In many cases, the expertise of traffic signal operations is found in one or two people, who are difficult to replace. This is especially true in smaller agencies. For this reason, collaboration across agencies may provide a critical mass of both infrastructure and staff required to effectively manage the system.

The benefits agencies can realize from participating in multi-agency collaborative efforts are adequately described in the FHWA document entitled: *The Collaborative Advantage: Realizing the Tangible Benefits of Regional Transportation Operations Collaboration*. This reference manual provides a six-step process to allow agencies to estimate their benefits of collaboration.

Regional collaboration offers significant benefits, and this report focuses on overcoming the barriers that hinder regional traffic signal operations programs. These barriers are not technological but rather

---

[1] Regional Transportation Operations Collaboration and Coordination – A Primer for Working Together to Improve Transportation Safety, Reliability, and Security, U.S. Department of Transportation, http://www.itsdocs.fhwa.dot.gov/jpodocs/repts_te/13686.html

institutional, organizational, and budgetary. Diminishing resources both hinder and necessitate the cohesiveness of traffic signal systems. However, specific examples of benefits include:

- Operating agencies increase access to funding by participating in joint funding applications.
- Agencies undertake larger, more technologically advanced projects by leveraging their expertise and resources with other agencies.
- Participating agencies help meet regional goals to reduce delay, fuel consumption, and emissions through coordinated initiatives, such as signal timing programs.
- Partners share communications assets to save resources and raise their collective ability to manage traffic on a regional level.
- Multi-agency collaboration has enabled the creation of joint dispatching that has resulted in decreased response time to requests for field assistance from partnering agencies.

Previous documents highlight the benefits of cross-jurisdictional coordination[2] and showcase regional strategies with case studies[3]. This document is intended to provide a framework for partner agencies to identify regional traffic signal management operational and maintenance policies and objectives and to establish regional priorities, activities, and performance measures that result in the achievement of those objectives.

---

2 Cross-Jurisdictional Signal Coordination Case Studies, U.S. Department of Transportation, February 2002, http://www.itsdocs.fhwa.dot.gov/jpodocs/repts_te/13613.html

3 Cross-Jurisdictional Signal Coordination In Phoenix And Seattle Lessons Learned from the Metropolitan Model Deployment Initiative (FHWA-OP-01-035) (2001), http://www.itsdocs.fhwa.dot.gov/jpodocs/repts_te/13222_files/13222.pdf

While there are many examples, the process within the Coordinated Freeway and Arterial Operations Handbook[4] will be used here because of the close link of arterial (and network) operations in general to the specific application of regional traffic signal management and operations. Regional Traffic Signal Operations puts the available elements together into an integrated package that maximizes system performance from the users' perspective.

Critical Elements Among Regional Stakeholders

- Resource integration, allocation, and management;
- Information documentation and exchange;
- Equipment sharing;
- Pooled funding;
- Personnel training, development, and integration;
- Systems integration; and,
- Institutional integration.

---

4 Coordinated Freeway and Arterial Operations Handbook, U.S. Department of Transportation, May 2006, http://tmcpfs.ops.fhwa.dot.gov/cfprojects/uploaded_files/06095.pdf

Integrating the Planning Process in the Development of a Regional Traffic Signal Operations Program

The federal aid transportation planning process is an important method to secure funding for management and operations, including traffic signal system. The *Transportation Planning Process Key Issues: A Briefing Book for Transportation Decisionmakers, Officials, and Staff* provides a framework for securing funding. This process is not one that can be tapped for immediate implementation, but is one that can be used, once its requirements are met, for ongoing funding.

The process centers on the Metropolitan Planning Organization (MPO) whose roles include:

Establish a setting: Establish and manage a fair and impartial setting for effective regional decision making in the metropolitan area.

Identify and evaluate alternative transportation improvement options: Use data and planning methods to generate and evaluate alternatives. Planning studies and evaluations are included in the Unified Planning Work Program or UPWP.

Prepare and maintain a Metropolitan Transportation Plan (MTP): Develop and update a long-range transportation plan for the metropolitan area covering a planning horizon of at least twenty years that fosters (1) mobility and access for people and goods, (2) efficient system performance and preservation, and (3) good quality of life.

Develop a TIP: Develop a short-range (four-year) program of transportation improvements based on the long-range transportation plan; the TIP should be designed to achieve the area's goals, using spending, regulating, operating, management, and financial tools.

Involve the public: Involve the general public and other affected constituencies in the four essential functions listed above.

> "A region is considered to be any multi-jurisdictional area defined by the collaborative partners."

The key issues are for the Regional Traffic Signal Operations Program to become part of the MTP and the TIP. The Regional Traffic Signal Operations Program needs to be included in the MTP in order to facilitate the funding of the program. FHWA has published *Management & Operations in the Metropolitan Transportation Plan: A Guidebook for Creating an Objectives-Driven, Performance-*

Based Approach -- Interim Draft[5]. This guidebook is designed to provide a basis on which to integrate transportation system management and operations into the metropolitan transportation planning process and to assist MPOs in meeting federal requirements under SAFETEA-LU. This calls for integration of management and operations strategies, including traffic signals, into the MTP. It highlights effective practices that result in an MTP with a more optimal mix of infrastructure and operational strategies, founded on the inclusion of measurable, performance-based regional operations objectives.

A Regional Traffic Signal Operations Program that builds on the regional planning process will build the institutional support and funding necessary for a successful program. This would be done by developing a Regional Concept of Transportation Operations, which is covered in **Regional Concept for Transportation Operations: The Blueprint for Action**[6]. A Regional Concept for Transportation Operations is a management tool to assist in planning and implementing management and operations strategies in a collaborative and sustained manner. Developing an RCTO helps partnering agencies think through and reach consensus on what they want to achieve in the next 3 to 5 years and how they are going to get there. A foundational concept presented in the RCTO is the need for developing SMART Objectives.

> An RCTO is a management tool used to assist in planning and implementing management and operations strategies in a collaborative and sustained manner.

SMART Objectives

Specific: It provides sufficient specificity to guide formulation of viable approaches to achieving the operations objective without dictating the approach.

Measurable: It is measurable in terms that are meaningful to the partners and users. Tracking progress against the operations objective provides feedback that enables the partners to assess the effectiveness of their actions. An operations objective is chosen that is measurable within the partners' means.

Agreed: Necessary for the development and implementation of the RCTO, partners come to a consensus on a common operations objective.

Realistic: The participants are reasonably confident that they can achieve this operations objective within resource limitations and institutional demands. Because this cannot be fully evaluated until the approach of the RCTO is defined, the partners may need to iteratively adjust the operations objective once the approach of the RCTO is determined.

Time-bound: Partners specify when the operations objective will be achieved. This promotes efficiency and accountability.

---

[5] Management & Operations in the Metropolitan Transportation Plan: A Guidebook for Creating an Objectives-Driven, Performance-Based Approach Interim Draft, U.S. Department of Transportation, http://ops.fhwa.dot.gov/publications/moguidebook/moguidebook.pdf

[6] Regional Concept for Transportation Operations: The Blueprint for Action, U.S. Department of Transportation, http://ops.fhwa.dot.gov/publications/rctoprimer/index.htm

The RCTO formalizes the collaboration and defines its direction for the future, essentially "getting everyone on the same page." By implementing an RCTO, partners put into action within 3 to 5 years operations strategies that will be sustained over the long term. The 3- to 5-year timeframe allows for many management and operations strategies to be implemented while keeping the RCTO tool responsive to current system performance needs. Additionally, the timeframe offers a middle ground between traffic signal operators who are focused on day-to-day activities and planners who are looking out 20 to 25 years.

Within any given region, there may be multiple RCTOs that focus on different operations functions or services. For the purposes of an RCTO, a region is considered to be any multi-jurisdictional area defined by the collaborative partners. That area may or may not coincide with the boundaries of a MPO. An RCTO focuses on operations objectives and strategies within one or more management and operations functions of regional significance, including traffic signals. The collaborating partners develop the RCTO to advance traffic signal management in their region and are driven by operations objectives that reflect regional expectations and opportunities. The partners may be motivated by a growing awareness of diminishing levels of service, a mandate from officials, or shortage of resources.

The goal of planning at the regional level is the development of consistent policies and objectives for traffic signal maintenance and operations throughout the region. Such activity developed and supported by key stakeholders in all affected agencies would be more likely to gain the approval of key decision makers responsible for funding. Once funding and resources have been harnessed, plans and procedures for specific activities can be developed. This may include resources to develop plans, document procedures, or implement projects on specific arterials. It may also include projects of a regional nature that would solve a host of issues globally for more than one agency, reducing repetition that would occur if agencies addressed the issues individually.

The RCTO provides a framework to connect operations to the transportation planning process offers benefits for planners who are interested in advancing cost-effective strategies to improve regional transportation system performance and operations-oriented partners who are seeking regional support for their joint efforts. An RCTO is one opportunity among several to link transportation planning and investment decision making to transportation system management and operations, as illustrated below.

Exhibit 1 Linking Planning to Operations[7]

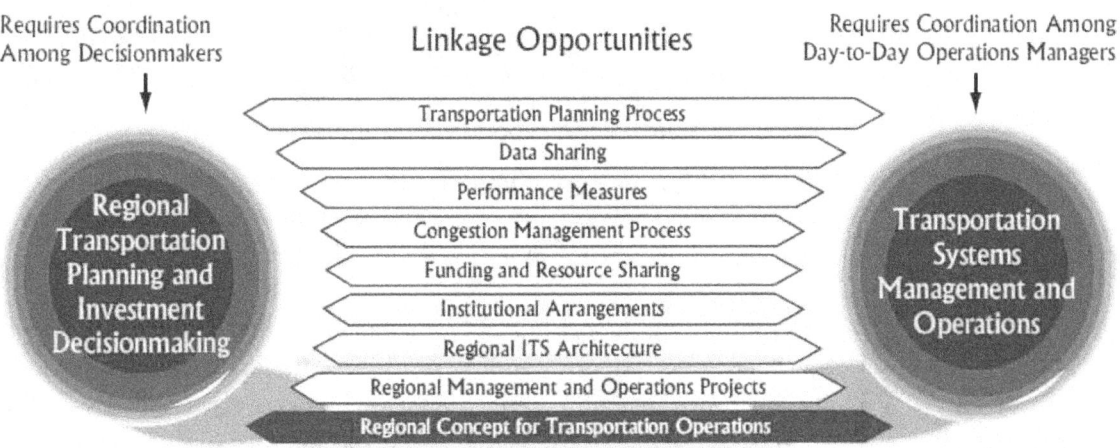

By linking to the planning process, partners can gain recognition within the region for operations and increase credibility with elected leaders whose support may be crucial in advancing operations. RCTO partners can ground their work in formally established regional needs, goals, and objectives. Additionally, they can increase the stability of their partnership by selecting the MPO to be an impartial and long-term host for the collaborative development and implementation of their RCTOs. RCTO partners may also be able to influence the selection of performance measures and data collection procedures used during regional planning to better track the progress toward the RCTO operations objective.

Funding for traffic signal system improvement, including management and operations, typically comes from local sources. Local funding sources are extremely competitive and traffic signal projects compete with virtually every type of capital project, including repaving, water and sewer projects, education, and public safety. Given the competitive funding environment at the local level, traffic signals do not typically fare well in the absence of a strong champion or transportation lobby.

---

[7] U.S. Department of Transportation, FHWA, Getting More By Working, Together Opportunities for Linking Planning and Operations, (Washington, DC, 2004).

State and federal funding for traffic signal management and operations is an alternative that is seldom pursued because it is perceived as a complex process and is riddled with institutional issues, such as the following:

- Management and operations needs are planned at the local level on a short-term basis of 0 to 3 years, while planners evaluate transportation infrastructure needs over a longer term of 20 to 30 years oriented towards major capital investment.
- There is difficulty demonstrating the benefits of traffic signal management and operations in terms that are understood by transportation planners and elected officials who approve the TIP, which feeds into the Statewide Transportation Improvement Program (STIP).
- There is limited understanding at the local level of the eligibility requirements for state and federal funds and how they are allocated and distributed.

Federal funding opportunities for management and operations do exist; however, state and local funding processes often make it difficult to allocate funds for traffic signal management and operations by creating separate categories of funds for capital and maintenance (operations is typically classified as a maintenance activity) expenses. The eligibility requirements and allocation of state funds is beyond the scope of this document as the process is not uniform across all states. Eligibility requirements for federal funds are outlined in Title 23 - Code of Federal Regulations. In TEA-21 and SAFETEA-LU, the Federal-aid Highway Program continued eligibility for federal funding of operating costs for traffic monitoring, management, and control systems from National Highway System and Surface Transportation Program funding. For projects located in air quality non-attainment and maintenance areas, and in accordance with the eligibility requirements of 23 USC 149(b), Congestion Mitigation and Air Quality Improvement Program funds may be used for operating costs for a 3-year period, so long as those systems measurably demonstrate reductions in emissions. Operating costs include labor costs, administrative costs, costs of utilities and rent, and other costs associated with the continuous operation of the system, such system maintenance costs,.

Traffic signal operations projects that are eligible for these other sources of funding must be included on the STIP. The STIP identifies the funding for, and scheduling of, transportation projects and programs and includes projects included in local Metropolitan Transportation Improvement Plans (MTIP). It is very unlikely that a single local jurisdiction's traffic signal management and operations project would be elevated for inclusion on the STIP. By taking a regional approach, a program may successfully elevate the status of these projects for inclusion on the STIP and hence federal funding. Examples of these funding sources are provided in Exhibits 2 and 3.

Exhibit 2 Federal Transportation Programs and Revenue Sources Administered by FHWA relevant to Regional Traffic Signal Management

Major Transportation Program	Federal Revenue Source
Interstate Maintenance	Highway Trust Fund with funds from federal: - Motor Fuel Tax - Truck and Trailer Tax - Tire Tax - Heavy Vehicle Use Tax
National Highway System	
Congestion Mitigation and Air Quality Improvement (in air quality non-attainment and maintenance areas)	
Surface Transportation Program (includes transportation enhancements and planning funds)	
Highway Safety Improvement Program	
High Priority (Demonstration) Projects	
Intelligent Transportation Systems	

Exhibit 3 Major Federal-Aid Highway Programs under SAFETEA-LU Eligible for Regional Traffic Signal Management

Program	Eligible Uses
Congestion Mitigation and Air Quality	A wide range of projects in air quality nonattainment and maintenance areas for ozone, carbon monoxide, and small particulate matter, which reduce transportation-related emissions
Interstate Maintenance	Resurfacing, restoring, and rehabilitating routes on the IHS, but no new capacity except HOV or auxiliary lanes in nonattainment areas
National Highway System (NHS)	Interstate routes, major urban and rural arterials, connectors to major intermodal facilities, national defense network. Fifty percent of NHS funds can be freely flexed to STP; 100% with US DOT approval
Surface Transportation Program (STP)	Broad range of surface transportation capital needs, including many roads, transit, sea, and airport access, vanpool, bike, and pedestrian facilities

The Metropolitan Planning Process

Metropolitan Transportation Planning and Programming requirements are outlined in the Federal Code of Regulations Title 23 Part 450. Each state is required to implement a continuing, comprehensive and intermodal statewide transportation planning process that includes the development of a STIP. The legislation mandates the formation of a MPO for each urbanized area within a state and that these organizations develop a TIP for their area. Consensus is built between the MPO(s), state DOT and regional stakeholders to reach agreement on the statewide plan, approved by the governor of each state. Eligibility for federal funding is contingent on a project's inclusion in the STIP. Some MPOs have included traffic signal timing and operations as line items in the TIP without designating funding levels. This allows an opportunity to allocate funds to these activities should delays or other factors free up funding from other capital projects.

Opening funding avenues for operations from sources such as the Congestion Mitigation and Air Quality (CMAQ) Improvement Program, Surface Transportation Program (STP), and state, regional, or local tax programs is a compelling reason to link regional traffic signal operations activities to the planning process. The ability of RCTO partners to

apply and receive funding in the near term depends on the flexibility of the planning organization to allocate funding for traffic signal management and operations projects. All projects need to be part of the MTP in order to be eligible for funding through the metropolitan planning process. In many regions, obtaining funding within one to two years is very difficult because all available funding is designated for specific projects many years in advance. In those cases, partners may choose to work to establish funding options for future management and operations projects while implementing an RCTO in the near term that relies on available resources and technology.

Examples

Pima Association of Governments (PAG)

The process used in the Regional Concept for Transportation Operations (RCTO) for the Tucson, Arizona metropolitan region include activities to identify the process described in the following section.

Example of RCTO – Pima Association of Governments

Project activities can be separated into two phases: Focus and Assessment, and Action Plans. This is illustrated in Exhibit 4.

The first phase, Focus and Assessment, includes the following steps:

- Establish Operations Objectives and Vision.
- Document Existing Polices, Practices, and Procedures and Existing Institutional Relationships.
- Establish Operations Goals.
- Establish Performance Measures.

The second phase, Planned Activities, includes the following:

- Develop Action Plans that are needed to Achieve Operational Goals. The Action Plans include the following:
    - Identify Necessary Institutional Arrangements and Relationships
    - Identify Resources Required for Implementation.
- Begin implementation of the Action Plans.
- Measure Operations against performance.

The following are the steps are used by PAG:

Step 1: Establish Operations Objectives (Mission and Vision for a Regional Concept for Operations)

The first step in the RCTO development process is to develop a vision that guides the remainder of the RCTO development process. The vision was developed through a brainstorming workshop with RCTO stakeholders. An important purpose of the Vision is to help other stakeholders, including agency executives and elected officials, to understand the purpose and nature of the PAG RCTO. In conjunction with the visioning process, a PAG RCTO Handout was prepared for distribution to agency executives and stakeholders. The Handout provides background and definition of the RCTO, discusses existing coordination and collaboration activities, and presents a Vision Statement for the PAG RCTO. The Handout material is presented in Chapter 3 of this report.

Exhibit 4   RCTO Project Activity Phases for the Pima Association of Governments

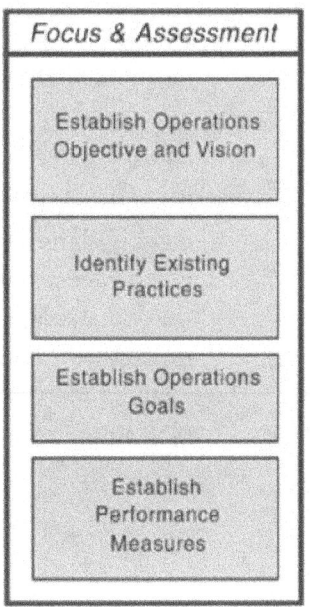
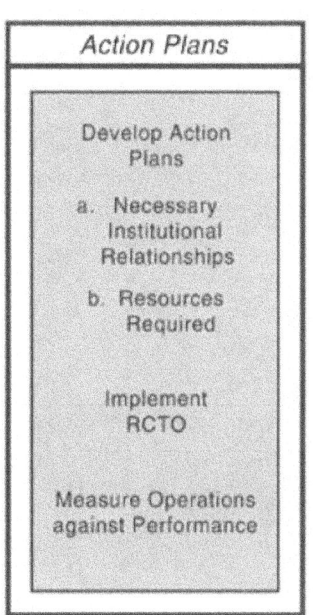

Step 2: Document Policies, Practices, Procedures and Existing Institutional Relationships in the Region

The second step in the RCTO development process is to conduct an inventory and compilation of existing transportation operations policies, procedures, practices, and resources of the member agencies as they relate to surface transportation system operations. This step focuses on the operational areas of traveler information, work zone management, arterial operations, freeway operations, and incident management. This task is instrumental to identifying the members of the "regional operations table" as well as establishing needs related to specific operations areas.

Step 3: Establish Operational Goals and Performance Measures

The process of defining Operations Objectives narrows the focus of the PAG RCTO to how arterial management, traveler information, work zone management and incident management need to be approached on a regional level and how they fit into the TIP process. In order to incorporate the TIP process into the RCTO, and mainstream operations into the traditional planning process, aspects of the Operations Objective extend into the 5 to 10 year range.

Step 4: Establish Performance Measures

Once the operations objectives are established, qualitative or quantitative performance measures are defined for each operations objective. The process of defining performance measures simultaneously with the operations objectives serve as a check that the operations objectives are specific and measurable utilizing existing or easily attainable data.

Step 5: Develop Action Plans to Achieve Operations Objectives

The next step in the RCTO development process is to develop specific Action Plans for each RCTO operations area. Action Plans identified new practices, or modifications to existing practices and procedures, that are needed to achieve the operations objectives. A desired outcome of developing Action Plans is that local agencies can modify some of their existing practices and activities to consider the regional operations goals and objectives.

Institutional Relationships: The Action Plans include discussion of the institutional relationships that are needed to achieve the operations objectives. Institutional arrangements are often the key component of an operations concept that determines the success or failure. While engineers are able to overcome most any technical challenge, developing sound relationships (including intergovernmental agreements) are many times the most challenging aspect of a project.

A primary consideration in identifying required institutional relationships is to recognize the relationships that are already in place and are working well. Stakeholders emphasize that they do not want to create new institutional layers or regional forums, as existing forums in the PAG region already provide a solid, established foundation for addressing regional transportation operational issues. As an example, a key component of the existing institutional framework already in place is the PAG Transportation Systems Subcommittee (TSS). This group brings decision-makers and practitioners together in forums that are dedicated to planning and operating the transportation system.

A second important consideration in the recommendation of institutional relationships is to capitalize on existing joint operating agreements. As an example, the Tucson region already jointly operates the traffic signal system. As another example, the Arizona Department of Transportation (ADOT) has established a multi-agency freeway management system with open coordination and contributions from several stakeholder groups. Furthermore, statewide and regional signal system procurement and Intelligent Transportation System (ITS) on-call consulting services contracts exist that allow multiple agencies to procure ITS equipment or consulting services through a common procurement effort.

Document Roles and Responsibilities: The Action Plans document specific roles and responsibilities for activities that are required to achieve the operations objectives. This includes general roles and responsibilities of local agencies and jurisdictions for planning, implementation, and operations activities related to the Action Plan.

Within each Action Plan, where feasible, a 'champion' structure is identified to guide the implementation of the Action Plan. As an example, the PAG Transportation Systems Subcommittee is identified to champion a Regional Traffic Signal Operations Program with oversight on the program's focus areas, implementation and funding. The subcommittee is guided by PAG staff and seeks to establish a working group to guide program activities and work out details associated with the program's coverage. While actual signal timing plans are not developed by PAG staff nor the subcommittee, staff and the working group are responsible for overseeing the administration of the program and working to 'elevate' the issue within the regional transportation community, and to facilitate the allocation of resources to develop and update signal timing plans. PAG staff or an identified individual from the working group update the TSS, perhaps on a bi-monthly basis, concerning recent activities relating to the Action Plan.

Resources Required: The PAG RCTO identifies the resources necessary to successfully implement each Action Plan. Identified resources range from physical improvements, such as TMC facilities, ITS infrastructure and equipment, or staffing to adequately operate the devices and field equipment. A critical element of this step is to identify operations-oriented projects that should be included in the PAG TIP or that can be pursued through the Regional Transportation Authority (RTA) Plan funded by regional sales tax dollars. The PAG TIP provides an opportunity to include specific projects that provide measurable and operational benefits. Although the TIP is on a five-year schedule, agencies need to begin submitting high priority operational projects for consideration. Stakeholder agencies have recognized that coming together to coordinate transportation operations has provided advantages to acquiring highly competitive TIP funds for Action Plan initiatives. Multi-agency programs, such as those identified through the RCTO process, are seen as high priorities across the region.

Step 6: RCTO Implementation Activities

Some elements of the RCTO Action Plans allow for immediate implementation. Others, particularly those that require significant funding, are implemented as funding becomes available. During the later part of the RCTO development effort, implementation began in the areas of Traveler Information and Work Zone Management.

Other Examples

The Denver Regional Council of Governments (DRCOG) established their program within the TIP to create funding pools for ITS or arterial traffic signal systems. Their program has allowed agencies to apply for and obtain funding in the near term for specific projects within those areas that have recently been defined.

In the Phoenix, Arizona area, a collaborative regional traffic signal optimization initiative in the region's RCTO was funded through the Maricopa Association of Governments (MAG) with CMAQ funds that became available for programming during the TIP closeout process. The success of the initiative caused the MPO to become very supportive of the regional signal timing program, and it will likely become a permanent part of the MPO work program. The Maricopa County Department of Transportation in Arizona provides Highway User Revenue Funds through its TIP to support an arterial incident management initiative outlined in the region's RCTO.

Additionally, the Hampton Roads, Virginia, region decided that more flexibility was needed in funding ITS and operations projects so it created a line item in the metropolitan transportation plan for these projects. Agencies can apply for CMAQ and regional STP funds for management and operations projects in the near term during the development of the TIP.

Summary

Operations issues that can be addressed and resolved at a regional level through policies, agreements, and plans include:

- Optimization and coordination of signals within and between agencies, such as state, local, county, and transit;
- Optimization and coordination of traffic signals with highway systems, such as interchange ramp termini and ramp metering;
- Altering arterial signal timing during freeway and arterial incidents;
- Sharing data on arterials during performance measure reporting, freeway and arterial incidents, and traveler advisories;
- Sharing maintenance practices and resources;
- Communication to both the public and elected officials with a consistent message;
- Documentation of practices and procedures; and,
- Training and development opportunities.

In addition to developing common regional-level operations policies, agreements, and plans, a traffic signal management and operations program could identify and address a number of other issues. Other tasks include the following:

- Establish different objectives and policies for varying arterial types, such as arterials within central business district or downtown, suburban, and rural areas. Objectives may vary depending on the type of land uses, travel patterns, travel speeds, and vehicle characteristics. An operations program should address these variations to maintain regional consistency.
- Establish a regional working group comprised of key stakeholders and a champion to lead the group in being responsible for traffic signal management and maintenance within the region.
- Develop the vision, goals, objectives, and performance measures for traffic signal management and operations in the region.
- Develop a regional traffic signal management concept of operations to identify high-level policies and plans needed to support plans and procedures for individual arterials. Such high-level policies should include:
    - Balancing major street throughput and average network/intersection delay;
    - Vehicle clearance times (yellow and all red);
    - Left-turn movement treatments (leading, lead-lag, lagging);

- - o Pedestrian treatments (rest in walk, leading walk, recall, clearance times, etc.);
  - o Signal timing monitoring and plan updates; and,
  - o Intersection hardware maintenance.
- Identify information and resource sharing needs on a regional level (e.g., identifying whether local agencies need to access and view freeway detector and closed circuit television [CCTV] cameras) for the purpose of traffic signal management and maintenance.
- Propose technology and ITS needs to support corridor traffic signal management and maintenance at a regional level.
- Assess and establish engineering and maintenance staffing needs and qualifications.

While a formal regional-level plan is desirable, it is not a requirement before developing plans and procedures for individual arterials. The benefits to developing a regional plan include:

- Resolving many of the institutional barriers on a regional level before proceeding to developing plans for individual arterials;
- Gaining region-wide consensus on the general approach and planning for traffic signal management and operations in the region;
- Gaining the support of key decision makers responsible for funding decisions in the region;
- Cost savings through addressing regional operations issues just once rather than multiple times when developing plans and procedures for individual arterials;
- Cost and time savings by early identification on what policies and agreements will be needed before proceeding to develop plans for individual arterials; and,
- Providing and maintaining a consistent message to the public.

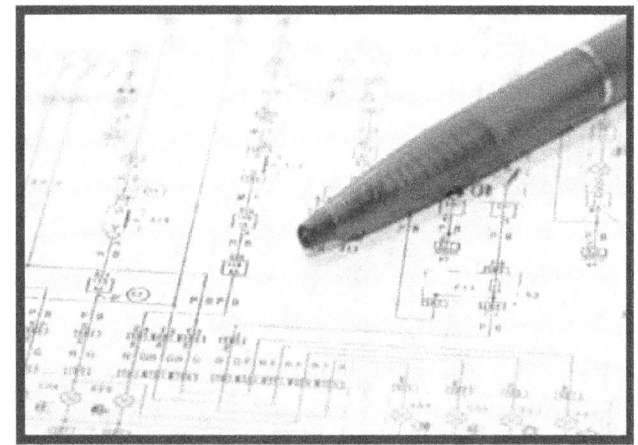

Some smaller regions may determine that the upfront cost of preparing a formal regional corridor management plan may outweigh the benefits listed above. For example, a region with one arterial may not need a formal regional traffic signal management and maintenance plan and could proceed directly to developing plans and procedures for that individual arterial.

After developing a regional traffic signal management plan, the region would have institutional structures in place, consensus by regional stakeholders, policies and plans to support traffic signal management and operations, a prioritized list of arterials warranting individual plans, and the resources identified to develop and operate individual arterial implementation plans. The next section discusses a recommended framework for development and operation of individual arterials.

Section 3
Common Threads

Common Threads

Regional Traffic Signal Operations Programs are unique in how they start, evolve, and sustain themselves. There is no one formula on how to start or develop a Regional Traffic Signal Operations Program. However, one common thread between successful programs is that stakeholders are committed to the program goals and objectives and have an interest in improving by cooperation. Without this commitment, a regional program cannot succeed.

Commitment to cooperation goes beyond contributing and participating. Commitment is likely the biggest hurdle for any Regional Traffic Signal Operations Program. This requires a change in approach and philosophy from each stakeholder. One element of getting everyone on the same agenda is through the development of "SMART" objectives[8]. Objectives are specific, measurable statements relating to the attainment of goals. Regional Traffic Signal Management objectives should be multi-jurisdictional in nature.

A review of successful programs indicates there are similar characteristics within the program's approach that has allowed stakeholders to commit. These characteristics include a culture of collaboration, informal partnerships, regional organizations, and regional resource allocation. Successful programs experience an evolution of these elements in different ways, but there are evident transitions of regional objectives and collaborations from an informal level to a structured collaboration and then to a mainstreamed program. The following sections describe each of the elements and how they are necessary to a successful regional transportation operations program.

> Similar to what occurs in the military related to joint exercises, collaboration and commitment to a regional program has to go beyond superficial efforts.

Creating a Culture of Cooperation

Success in a regional program is aided by a culture where stakeholders are committed to working together for mutually beneficial transportation operations. The necessary culture is likely to necessitate change, and a change to any culture is challenging and takes time. The culture is created by champions who see issues that are common across agencies and seek to reach across agency boundaries to improve.

> The FHWA Traffic Signal Operations Peer Review Program has been instrumental in facilitating a cultural shift, helping agencies that recognize there has not been a common voice by which to request additional attention to traffic signal operations issues.

8 Management & Operations in the Metropolitan Transportation Plan: A Guidebook for Creating an Objectives-Driven, Performance-Based Approach, Federal Highway Administration, FHWA-HOP-08-007, November 2007.

Champions attempt to bring various stakeholders together to begin the dialogue to assess the current conditions and ways they might make effective change. The necessary change often starts with a single person who champions the importance of regional collaboration. These individuals develop support from their agency, especially among the public officials and other top level leaders. This might be done by having a peer review or signal operations audit completed to respond to criticism or proactively to meet a specific new challenge. Participants must work together to define opportunities for common objectives

that will allow coordination as part of the practice. As cooperation opportunities are considered in projects, visible benefits can be demonstrated to other agencies that will help grow the network. As the practice continues, more people will witness the benefits and become mainstreamed into additional activities leading to further partnerships that begin the basis of regional efforts.

Developing Successful Partnerships

Informal partnerships have proven to be integral to developing a collaborative environment and a base of experience for working together that creates a sustaining force behind the program. As the relationships grow, the partnerships may result in additional collaborative efforts such as joint studies into new technology or something similar. Finding a champion that can create this sort of partnership requires participation within either an informal arrangement or formal agreement between agencies. Thoughtful guidance of these collaborative efforts is essential, and as relationships develop the involved parties should agree on mutually beneficial results. The evolutionary transition occurs when relationships transcend individuals and entire agencies become part of the process. Shared goals and practices are developed when transportation agencies assert themselves as leaders in the informal partnerships. As these practices and procedures are followed, a common voice is more easily understood by elected leaders and appointed officials.

> Oftentimes, partnerships begin by regular conversations or meetings to compare experience with new technology, discuss sharing communications infrastructure, or joint procurement of equipment. The Denver Regional Council of Governments recently completed a Transit Signal Priority Study to insure consistency across agencies.

The key element of partnerships that propel a region to transition from an informal process into an integrated and streamlined program is effective leadership and management. The presence of technical or political leaders that act as a catalyst for change must have a strong message for success. The importance of having a common voice within an agency and the region allows a region to share resources, identify goals, and coordinate communications internally and externally. A leader can leverage common efforts to communicate a region's goals and expectations to the decision makers as well as the public. The ability to provide and notify the public that the region is unified on a common approach improves public perception.

Casting a vision that allows regional partners to see what's in it for them and rallying their support is a key aspect of regional leadership and management. If it is a regional program, support comes from the agencies within the region that sponsor the concept of operations. This is accomplished through interagency agreements where multiple agencies share in responsibilities or through less formal agreements between agencies. Without these agreements (formal or informal), individual agencies can be significant hindrances in regional efforts.

Examples of Collaborative Activities

Several regions allow one agency to take the lead in signal system management that permits sharing technical resources and mutually beneficial outcomes. The **Portland-metropolitan area** has benefited from efforts to share development costs of a traffic signal system. The City of Portland led the selection process for procuring the equipment, but it has involved stakeholders from neighboring communities so that these agencies could utilize a common platform. The cooperation created by this process has also led to joint efforts to agree on a new traffic controller standard (including software) and selection of an adaptive control algorithm. The Oregon Department of Transportation is following the City of Portland's lead with the joint efforts of having a compatible signal system state-wide.

The **Mid-America Regional Council (MARC)** is the association of governments for the greater Kansas City area in Kansas. MARC's goal is to build a strong regional community through interagency coordination, cooperation, planning, and leadership. MARC serves the role of a forum that allows member agencies to address regional objectives and issues as well as providing technical resources for the region. Within the roles and responsibilities, MARC serves as the keeper for the long-range transportation plan, regional intelligent transportation system architecture, and the congestion management system.

The **Southeast Michigan Council of Governments (SEMCOG)** serves as a leader and management for the greater Detroit metropolitan area. The SEMCOG includes numerous cities, seven counties, and educational centers including the University of Michigan, Ann Arbor. This MPO is a member organization that advocates for improvements that benefit the greater community and provides consistency on general transportation practices, including traffic engineering practices. Many of the policy decisions are made by local agencies within the SEMCOG's member governments, which allows for this consistency. Additionally, the SEMCOG also serves as a database for current and past regional projects so that any member agency can utilize as guidance and education.

Evolving the Organizational Structure

The evolution of a regional program begins with the self-assessment and recognition of ways to improve by working across traditional boundaries. When an organization realizes that they have a problem (i.e., understaffed or with limited technical knowledge), the agency may seek efficiencies. The evolution can be initiated by the recognition of the need for more training that they need to combine with adjacent communities to organize the requisite classes for staff development. An organization that has a regional focus may be able to coordinate needs across multiple agencies.

> **Example of Collaborative Training**
>
> The Los Angeles County Metropolitan Transportation Authority (Metro) has developed a training program that provides both local agency staff and outside professionals an opportunity for professional development. The program was initiated after officials recognized the need to provide on-going training for the local agencies to understand the design and maintenance of the continually changing signal system. Los Angeles Department of Transportation and consultants were contracted by Metro to develop and deliver the training. The program assisted internal staff to provide consistent quality of work across the region as well as educating staff on different technological capabilities. The program grew to include consultants and outside professionals, which also allowed the outsourced projects to become more consistent to Metro's standards.

The Regional Concept for Transportation Operations Primer suggests that an organization begin by tackling a relatively simple operations issue, such as traffic signal timing, in which the need for improvement is widely acknowledged and easily understood by several agencies. By identifying a common need, the agencies can share access to resources and increase the need for a structure that allows information sharing and communication between agencies.

The organization structure may transition from a group with ad hoc unplanned activities to one with more scheduled activity and regional collaboration as needed. This transition occurs at varying levels and at different paces, depending on the projects and amount of sharing that occurs. Multi-agency projects result in an evolutionary change that makes regional programs a success.

Section 4
Framework for a Regional Traffic Signal Operations Program

Framework for a Regional Traffic Signal Operations Program

A process roadmap, or framework, is needed to facilitate the life cycle (planning, design, operations, and maintenance) of traffic signals for a region. This framework is recommended for use after addressing regional-level issues identified above. Because regional traffic signal management is typically fragmented due to the institutional makeup of the agencies involved in traffic signal management and maintenance, the framework provides a process to overcome the institutional seams that inhibit coordination and collaboration.

The framework described in Exhibit 5 is scalable based on the complexity of the system and required operations strategies. The framework focuses on the continuous nature of a Regional Traffic Signal Operations Program. The element "steps" in the process are intended to facilitate addressing all the issues in a successful program. For some programs, some steps may not need to be formally addressed (e.g., a formal evaluation of operations strategies may not be necessary when there is only one feasible strategy). For larger, more complex networks, formally going through each step in the framework may be desirable to ensure consensus is achieved by stakeholders, to provide a roadmap through the entire lifecycle of the project, and to help identify and address major problems before it is too late (i.e., already in the implementation phase).

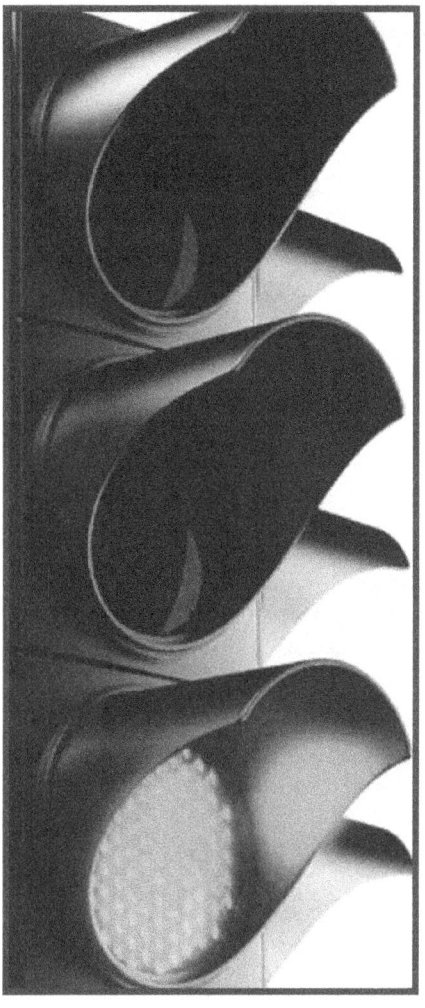

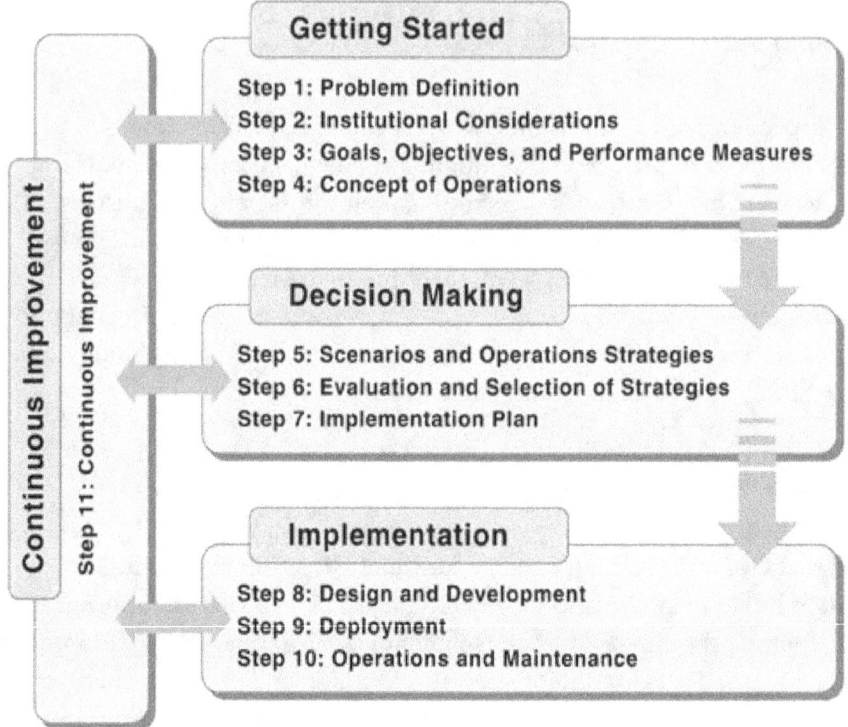

Exhibit 5 Regional Traffic Signal Management and Operation Program Framework

It should be noted that the framework is cyclic because the process often requires iteration between steps to resolve competing issues. For example, an operations strategy may be chosen for evaluation. Upon close consideration, the strategy may require more resources than are available, resulting in the reconsideration of alternative strategies more consistent with the available resources.

The steps in the framework are grouped into four categories: getting started, decision making, implementation, and continuous improvement. The remainder of this section presents a high-level discussion of each of these four categories. The next section will look at sustaining the program, including drawing on experiences from successful programs.

Getting Started

Step 1 is to define the problem. The definition of the problem in the broadest sense should begin at the regional planning and coordination stage discussed in the previous section. At the regional level, problems are identified at a minimal level of detail as part of a planning and programming function. What may be a problem to one agency may not be a problem to another; therefore, it is important to discuss and gain consensus on the extent and severity of the problem to be addressed. An example of a high-level problem is necessary stops due to the lack of signal coordination between jurisdictions.

Step 2 is to assess institutional considerations. A framework can create a bridge for agencies to address an identified problem in a manner that all affected stakeholders can support. Part of overcoming institutional isolation is establishing a structure, such as a working group composed of various representatives from the affected agencies, early on to guide the entire process. The structure; whether an ad hoc group or formal committee should consist of as many of the regional stakeholders as possible. Participants should be representative of all agencies and parties involved in the planning, design, operation, and maintenance of the plans and procedures. The size of the group should be commensurate with the size and complexity of the project. Overall, the success of the project is a direct result of the ability of a variety of institutions, agencies, and affected parties to gain consensus, have regular contact through meetings or other communication, and work within the context of other regional entities. In the case of cross-jurisdictional signal timing, the process could be relatively simple when there are few plans and coordination is achieved through time-based coordination. If the concept of operations (described below) includes interoperability between agencies, then the complexity is significantly higher.

Once the structure for the stakeholders has been agreed upon and developed, it will be their responsibility to identify the goals, objectives, and performance criteria for the project, which is Step 3 in the process. Ideally, the goals and objectives established will be directly related to the problems identified in the first step. If the problem was identified broadly, it will be the responsibility of the stakeholder group to examine the problem and develop a more detailed assessment of its nature. The goals will be broad statements of the desired outcome (for example, reduced travel time) once the problem is resolved. The objectives will be specific statements of what will be achieved in support of the goals (e.g., to reduce transit-related delays by 10 percent), and the performance measures identified will represent specific measurements that will be used to assess the goals and objectives (e.g., person-hours of delay during the a.m. peak travel period). Performance measures should include metrics that users of the transportation system experience directly, such as travel time between points.

Step 4 in getting started is to develop a regional concept of operations. The concept of operations is a document, either formal or informal, that provides a high-level, user-oriented view of operations on a specific arterial or network. It is developed to help communicate this view to other stakeholders in this process, such as the interested public, and to solicit their feedback. The final document should describe the goals, objectives, and performance measures of the project agreed upon by the stakeholder group. It should also provide a description of the existing conditions, operating practices and policies, and system capabilities. The document should include a description at a high level; the operational scenarios (the traffic conditions during a given operational deficiency) when strategies, plans, and procedures are needed; and the high-level strategies that can address the problems during these scenarios. Again, complexity may be minimal in simple projects (e.g., time-based coordination) and significant in complex projects (e.g., transit signal priority).

Decision Making

Decision making starts with Step 5 and requires identification of specific operations scenarios for the arterial, which allows the process to move toward the selection of strategies. While an example of a high-level scenario may be to reduce transit traffic signal delay, a more detailed scenario would be to implement transit priority on Main Street in downtown during morning and evening peak-periods. Based on this scenario, a number of operations strategies might be identified, including both active priority at intersections with excess capacity and passive priority at intersections at capacity. Examples of the types of strategies that might be appropriate within a regional context include:

- Consistency in signal timing practices (i.e. clearance intervals, intersection configuration, pedestrian timing and policies);
- Reporting and responding to citizen complaints and providing traveler information;
- Identification of regional priorities, corridors of significance, performance goals, performance measurement;
- Outreach to the public and decision makers;
- Cross jurisdictional timing;
- Region-wide transit signal priority;
- Implementation of incident management plans;
- Implementation of severe weather plans; and,
- Adaptive traffic signal control.

While several strategies may be easily identified, Step 6 is to evaluate and select the most appropriate strategies for each scenario. The assessment of strategies can vary from simple, pragmatic assessments (e.g., the coordination across jurisdictions on major arterials may not warrant significant analysis) to detailed traffic micro-simulation studies that may be necessary to assess complex transit signal priority options. The method used should be appropriate to the complexity of the alternatives and the cost of implementation. The evaluation criteria should also be representative of the goals and objectives of the local area.

Example of Regional Cooperation

The Puget Sound Regional Council formed a Regional Traffic Operations Committee following the FHWA peer review focused on providing regional management of traffic operations. It has recently published a Best Practices document and is working on an RCTO and a Regional ITS Implementation Plan.

Outcomes of the FHWA Peer Review

The Portland (OR) metropolitan region's MPO, Metro has had a long standing ITS subcommittee to provide guidance on technical issues, but never had a long range plan for ITS integrated into the regional transportation plan. Following the FHWA peer review, the City sought grant money to develop a long range Transportation System Management and Operations (TSMO) Plan to develop a list of collaborative projects to improve its effectiveness within the planning processes.

The final step in the decision making phase of the framework, Step 7, is to develop an implementation plan, which provides all the information necessary to proceed with the implementation phase. The project implementation plan may be a formal document that represents the sum total of all the work that has been undertaken up to this point, summarizing the identified problems; the goals, objectives, and performance measures; the corridor concept of operations; and the selected strategies for various scenarios. When describing operations strategies, details such as capital and operating costs, potential programming priorities and scheduling, infrastructure needs in support of the strategies, and descriptions of maintenance procedures and costs can help create a valuable blueprint for the remainder of the project.

Implementation

The implementation phase consists of design and development, deployment, and operations and maintenance. Step 8, design and development, translates each of the various projects included in the implementation plan into executable project plans. Operations plans would be developed only at a high level in the implementation plan. For example, the overall implementation plan may identify an active transit signal priority as an operations strategy, and the plan would identify the necessary hardware and software (on both the transit and the traffic side of the system). In the design and development stage, the exact traffic signal settings in the timing plan would be determined, the specific individual(s) authorized to implement the plan would be identified, and details of how the plans would be implemented would be agreed upon. The design of needed timing plans to support the operations strategies would also be completed at this stage of the process. Any needed interagency operating agreements would also be finalized during this step.

Step 9, deployment, comprises signing interagency agreements and the implementation of the new strategy, or "turning on" needed software or communications equipment. There could be many factors involved with a multi-agency, multifaceted deployment, so patience and dedication is necessary to ensure a successful project.

Step 10, operations and maintenance, is perhaps the most important in the process. Stakeholders whose primary responsibility will be to activate and operate the system should be involved early in the process of developing operating plans and procedures. These individuals will not only provide valuable insight into operations and management processes, but also have a clear understanding of the roles and responsibilities of operations and maintenance personnel during the various scenarios. They will also be able to ensure that plans are operating at maximum efficiency and reliability.

Continuous Improvement

The continuous improvement process, reflected in step 11, is never ending. As the system is being operated and maintained, it must be continually monitored and evaluated. The monitoring process determines whether the actual performance of the system matches the goals and objectives of the project. If the system is not solving the problem identified in Step 1 or meeting the performance metrics identified in Step 3, then modifications should be made to better address the problem. This is, in effect, another cycle of problem identification and identification of improvement strategies, evaluation, prioritization, design, deployment, operations, maintenance, and so on. Without such a process, the project (and the overall system) will fail to perform at optimum effectiveness and efficiency.

Summary

This section presented a broad view of the planning-level activities recommended for successfully developing and sustaining a Regional Traffic Signal Operations Program. The planning process was examined at a regional level and then at a project (arterial or network) level. It is only after understanding these levels that meaningful analysis can be initiated. The framework introduced in this section is described in more detail in *A Coordinated Freeway and Arterial Handbook*. Utilizing a consistent framework insures a repeatable, stepwise process. The next section will look at key elements to developing and maintaining a sustainable Regional Traffic Signal Operations Program.

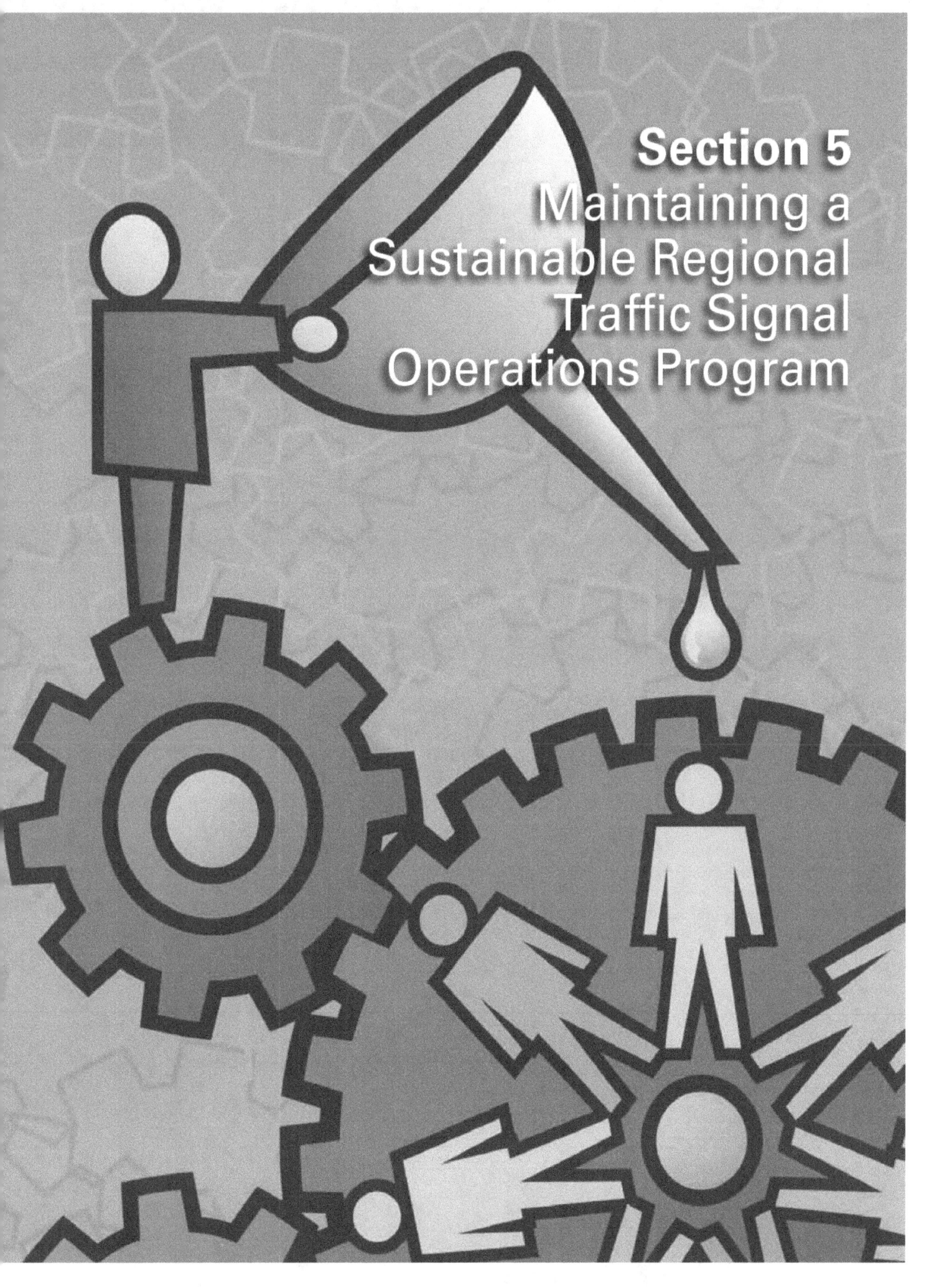

Maintaining a Sustainable Regional Traffic Signal Operations Program

This section describes the key activities necessary to maintain a successful Regional Traffic Signal Operations Program. Similar to the common threads identified previously, these elements are directly related to a human element of communication and commitment. People initiate the necessary actions to implement regional collaboration, and it is communication that maintains and nurtures committed relationships. The communication can be characterized as staff to staff, agency to agency, agency to regional oversight, and agency to public. The following five elements are necessary for maintaining and sustaining a Regional Traffic Signal Operations Program:

The first element is leadership and management. Regional champions carry the burden of initiating collaboration and a change in the culture. The relationships and trust need to be in place to allow regional programs to succeed. Once leadership and management is developed, both technical and personnel support is required. Technical support comes from identifying, monitoring, and reporting performance. Performance measures provide the technical support for the collaborative efforts as direct correlations can be made between regional projects and operational benefits. The other support comes from qualified staff members who need technical training and professional development. With trained internal and external professionals, traffic signal operations can be maintained at the desired performance levels. Support for these projects and staffing requires funding. There are many funding sources and mechanisms in place to help support programs; however, money is a limited resource, and there is usually strong competition with other types of projects. Since funding can be challenging, there is a need for the regional transportation community to gain support from the public and elected officials. Outreach helps educate others by letting them know what projects a program is doing and what is being accomplished. The following sections describe further the details of each of the identified elements and how each helps in developing and maintaining a Regional Traffic Signal Operations Program.

Keys to a Sustainable Program

- Leadership
- Self Assessment and Evaluation
- Performance Measures
- Training Program
- Funding Mechanisms
- Public Involvement and Outreach

Leadership

It has been well documented that leadership is the most important key to a successful program. It often starts with a champion that takes it upon his/her shoulders to motivate and engage others. The leadership role often starts with a single person but can evolve into becoming an organization where a region shares a common voice. An organization or department with an agency, such as the MPO, may be the starting point for regional collaboration. The design of a MPO specifies how regional goals are subdivided and reflected for the parties involved in the long-range planning. Finding other stakeholder champions to participate in a MPO committee or even within the MPO will help maintain leadership within the group, but also expand into leadership within the various agencies.

Best Practices—Leadership

The **Puget Sound Regional Council (PSRC) in Washington State** is leading the regional efforts to identify smarter ways to move people and goods. The PSRC's Regional Traffic Operations Committee has identified regional signal coordination as the functional area of focus for future activities. It also recognizes that investment in traffic signal operations as well as integration of ITS devices results in a high benefit to cost ratio. As a result, PSRC's incorporated both highway and arterial operations into their long-range transportation plan, Transportation 2040.

The **Pima Association of Governments (PAG) in Tucson, Arizona,** developed an ITS architecture that outlined traffic signal operations for the region and how to incorporate technology to improve the ability of the region to share information with their elected leaders. Planners in PAG utilized the RCTO planning process to define regional needs and their commitment to achieving specific goals. By defining these goals, PAG led the way for agencies to work collaboratively and implement ITS effectively for the agency's operational goals.

Self-Assessment and Evaluation

Self-assessment is a process by which you learn more about yourself or agency – what works, what could be improved, and what keeps an agency operating well. There are several existing resources available for agencies and regional organizations to assess their traffic signal operations opportunities and to enhance the efficiency and effectiveness of regional practices. The first resource is the ITE traffic signal operation self-assessment and the *Traffic Signal Audit Guide*. The second resource is FHWA. FHWA offers a peer review process and a multitude of existing documents that are available for all professionals.

Results from both the 2005 and 2007 National Report Card have raised the national awareness of the state of practice for traffic signal operations. The ITE traffic signal operation self-assessment was updated in 2007, reflecting comments and feedback from the original assessment in 2005. The traffic signal operation self-assessment is intended for an agency to evaluate its traffic signal operations and maintenance. There are four main benefits of the self-assessment as identified by ITE and include:

- Increasing national awareness of the need for improved traffic signal operation;
- Increasing local awareness of the need for improved traffic signal operation;
- Identifying strengths and opportunities; and,
- Providing a benchmark for performance.

These benefits of the self-assessment emphasize the need for both national and local awareness for traffic signal operation as well as identifying what areas agencies need to improve. The "score" or rating provides agencies an indicator of overall performance. The self-assessment identifies strengths in a traffic signal system and the various opportunities for improvement. Additionally, the evaluation process serves as an education tool for agency staff. The self-assessment also determines a benchmark for traffic signal operation practice. The various questions include an example of good practices where staff can compare their own practice and then learn how to improve.

Another resource for regional organizations and agencies is a peer review process. The purpose for a peer review is to provide an evaluation of system management and operations practices from a team of objective experts that are familiar with national best practices. Outcomes of this process provide the region with a comprehensive review and a series of recommendations for improvement. The reviews typically begin with a discussion of what and/or how the self-assessment rates the region. This is used as a starting point on the workshop participants to receive constructive criticism and feedback. Other discussion items include regional management and operations, signal timing practices, coordinated systems, data collection and archiving, maintenance, partnerships training and professional development.

Best Practices—Self Assessment

In Arlington County, Virginia, public works groups are evaluated on the number of public comments per year. It is each group's goal to reduce the number of comments/complaints from the previous year. For the traffic operations group, the comments range from maintenance issues to signal timing. Each comment requires action by County staff members and at a reasonable response time. A complaint metric is one performance measure that must be carefully balanced against the agency's signal timing operations objectives.

FHWA staff have partaken in several workshops nationwide including Ada County, Idaho; Portland, Oregon; Puget Sound, Washington; Washington, DC; Puerto Rico; Fargo-Moorhead, North Dakota; and Minneapolis, Minnesota. Outcomes of the workshops have varied, but one common theme has been that each region has been energized with the guidance and advice received. Along with receiving feedback for the region, the workshop encourages collaboration amongst traffic signal operation professionals.

Performance Measures

Regional programs often utilize measures to identify the need for increased funding. Performance measures that are quantitative in nature can be used within projects to measure progress and assess how a strategy is working. It is common to link benefits with strategies to highlight the change in measures. For instance, the Washington Metropolitan Council of Governments uses the number of signals retimed to identify progress toward their goals. Based on this measure, the agencies can report on performance regarding traffic signal improvements.

The first step is to recognize the need and the benefits of having a performance metric for the region. This is often a challenge as there are many metrics and each holds its own importance. A key element of establishing this is to utilize a metric that is within an agency's means, including budget, staff, and equipment. An agency must know its available resources and limitations prior to defining metrics. The second step is monitoring the traffic signal system to determine the need for signal timing adjustments or maintenance. Once a system or corridor metrics are beyond the ability of basic maintenance, then corrective measures should be identified. This may include a complete reevaluation of the system.

There are many ways agencies can measure traffic signal operation performance, including average corridor travel time, average delay, and how often a signal/corridor is retimed. Additionally, there might also be consideration of multimodal objectives with special attention paid to transit or pedestrian operations. Having a regional method of evaluating signal systems can aid in determining areas for improvement, managing resources, and finding funding sources. Once there is a defined set of performance measures, it is important to ensure these support the goals and are monitored. A common challenge occurs when complaints arise at an intersection that had recently been updated, and the agency adjusts the timing to accommodate the complaint. While flexibility is important, the actions must consider the compromises that result as a part of the change.

> **Best Practices—Performance Measures**
>
> The Maryland State Highway Administration (SHA) has created a business-style model for maintaining their traffic signal operations, setting yearly goals for the number of retimed traffic signals and corridors. SHA's focus objective is "mobility" across the transportation system. This allows for a focus on reducing delay on the state arterials. Additionally, it quantifies environmental measures such as reduction in fuel consumption and emissions, arterial through-put, and evaluating crash histories along the corridor. Currently, SHA sets an annual goal of retiming 400 intersections to reduce delay. With this annual goal, SHA can gauge how effective it was on completing the scheduled corridors. SHA also realize that the goal is relative and cannot always be met since the percent of delay reduction will have diminishing returns as corridors continue to be retimed over time.
>
> Due to the diminishing returns, a future goal for the agency is to collect data on select corridors each year within the state to monitor the change in traffic patterns and volumes in order to better prioritize resource allocation to areas of significant growth. With this type of data, SHA staff will be able to determine if a corridor or intersection in the area requires updates. Understanding that it would not make sense to update signal timing in an area if there has not been much change, SHA could reallocate resources to other corridors.

As observed in the FHWA peer assessments, few agencies define performance metrics and fewer evaluate them. Often, agencies are in a reactive mode addressing public complaints instead of being proactive. In a recent poll, over 70 percent of practitioners made signal timing changes only when there are complaints. Some agencies use the number of phone calls as a performance metric. It should be realized that public complaints are an indication of failures, due to engineering, maintenance, or communication. Complaints are valuable information and are worth measuring, but it should not be the sole performance metric. Some agencies utilize the public as an extension of their staff as a source of information while other agencies utilize the complaint as an opportunity to involve the public as part of an education exercise.

> **Best Practices—Performance Measures**
>
> The Regional Transportation Commission (RTC) of Washoe County, Nevada, utilizes innovative public outreach relating the work it does to the public. It has utilized a website (www.rtcwashoe.com) to solicit feedback and to disseminate information about projects. The website is also used as one of the opportunities for a citizen to report a problem. The website also includes sections that explain the performance metrics and a video that describes the before and after of signal retiming. The County's PIO has worked with local media outlets to raise the awareness of signal timing and the associated benefits.
>
> A few agencies utilize a complaint as an opportunity to engage the citizens. It is an opportunity to educate them as well as to have them be part of the solution. When a complaint is received, agency staff takes time to discuss with the citizen about the concern or issue. During the discussion, the staff member asks if the citizen would be willing to partake in data collection to help the agency document it. Some agencies provide a GPS unit to citizens and allow them to travel with the unit over a period of time. The GPS data is then analyzed and reviewed with the citizen. If the complaint is valid, then the citizen is commended on bringing the concern to the agency's attention.

It should also be understood that an agency's performance measures can vary and could be based on the specific context of a community. A corridor may be a primary commuter route with heavy peak directional flow, and an agency may focus on throughput and reduction of number of delay. Application of a similar approach in a downtown environment is not likely to have beneficial effects, and different performance measures should be considered. Performance metrics, as with other regional policies, should consider the variations within their transportation network and have objectives applied accordingly.

Training Programs

Training is necessary for consistency in operations, providing the confidence and trust of other partners in the capabilities of their peers. Therefore, staff development is a key element in developing a regional program. Technical and leadership growth in staff allows agencies to become more confident in their abilities and also trust their regional peers. Shortage of qualified people is a significant challenge and has been linked to budget issues that have resulted in the inability to hire new staff but also to retain existing qualified people. This is not just a public sector issue, but is also of concern in the private sector as contractors perform a significant portion of the work. A well trained staff is the heart of all activities, and their professional growth is dependent on exposure to various training opportunities, both internal and external. Training opportunities provide many benefits to the agency as well as the employee. Such benefits include:

- Increased efficiency and consistency;
- Increased ability to learn and apply new technologies and methods;
- Increased employee satisfaction and motivation; and,
- Increased creativity and innovation.

There are many regional and national training programs that provide opportunities for traffic signal operators and engineers.

Best Practices—Training

The Los Angeles County Metropolitan Transportation Authority (Metro) has developed a training program that provides both local agency staff and outside professionals an opportunity for professional development. The program was initiated after officials recognized the need to provide on-going training for the local agencies to understand the design and maintenance of the continually changing signal system. The program assisted internal staff to provide consistent quality of work across the region as well as educating staff on different technological capabilities. The program grew to include consultants and outside professionals, which also allowed the outsourced projects to become more consistent to Metro's standards.

Best Practices—Training

Mobile Hands On Signal Timing (MOST) (www.webs1.uidaho.edu/most/) was developed with a federal grant by the University of Idaho for the FHWA to provide both students and professionals skills to use traffic data, signal equipment, software programs, and resource materials for traffic signal implementation. The MOST program integrates traffic signal controllers with simulation packages to allow the participants to develop, implement, and test traffic signal timing strategies. The simulation environment provides immediate results of the signal timing parameters that are programmed in the controller.

Another resource for external training opportunities is through professional organizations such as the Institute of Transportation Engineers (ITE). ITE is a resource center for the profession that provides opportunities to both learn and share. Publications, web-seminars, and conferences are a sample of opportunities that are available.

In addition to engineers, maintenance staff members also require training and development. It is common that staff learn "on the job," but there is a limit of knowledge based on the experiences and the surrounding staff. There are opportunities for staff to be trained and certified, which would provide a consistent knowledge base. This opportunity is through the International Municipal Signal Association (IMSA). The purpose of IMSA is similar to ITE; it strives to keep professionals trained and educated on up-to-date construction and maintenance procedures as well as provide information on new products and innovative developments in signal systems.

> Similar to the LA County Metro training program, the City of Fargo, North Dakota, also provides a training program and process for its maintenance staff. Staff are encouraged to be trained and certified at the various IMSA levels. However, the training does not stop with the certification process; the technicians developed an internal training program to review signal system procedures.

As indicated, successful traffic signal management and operations programs are based on people and their ability to communicate. The program participants require commitment, but also need to be motivated and trained. To achieve this, a consistent message of performance objectives, collaborative efforts, and professional development is needed. This begins with the leaders but extends to all support staff members. Staff members are the ones that design, maintain, and fix traffic signal operations based on the policies and objectives of the region. To ensure that the outcomes are consistent with the objectives, proper training is required.

Funding Mechanisms

Funding mechanisms identify the amount of resources, staff time, and equipment that can be applied in the collaborative effort to sustain the program. Investments in regional programs may involve preservation of existing funds, cost savings associated with efficiencies, partner agency budget allocations and commitments of staffing, equipment, or facilities to support regionally significant activities. The following section discusses the need for funding as well as innovative ways to find or preserve funding.

Traffic signal management and operations requires dedicated funding for both the engineering and maintenance. Municipal budget constraints create competition among different programs for funding. Capital projects may compete with operations and management (O&M) budgets in communities, and if sacrificed the quality of service provided to citizens suffers. Traffic signal system O&M is especially neglected because practitioners have limited tools to describe the effects of reduced attention. Degradation is less discernible to the traveling public because of the pace at which it occurs. Cutting O&M funding and staff is a short-term solution because the costs are seldom captured. As a result, the communities are likely to experience increased public scrutiny about the stewardship of the existing resources and decreases in services.

Budget constraints also impact strategic planning for improving the infrastructure and activities that may reduce redundancy or wasted time. For example, the following technology advances present opportunities to reduce workload:

- Communication to a remote signal may reduce the need to send a technician to address a complaint.
- Closed circuit television (CCTV) cameras may make a field visit unnecessary.
- Signal timings may be uploaded and downloaded from a Traffic Management Center thus reducing time in the field.
- System alarms can be directed to cell phones, email, or other remote communication devices to improve failure response times.
- Equipment upgrades may improve detector reliability and performance.

The workload savings associated with the above examples can be used to improve O&M. This in turn can lead to lower costs to the traveling public, increased preventative maintenance, and optimally timed signals that reduce congestion and delay.

A project developed through regional TIPs is often proposed by the regional agency in future year plans. Regional agencies work together with their public officials/decision makers to identify specific projects in the near and long-term future. The signal operation projects can be specifically for corridor timing updates, but can also be a part of larger improvement projects. An example of this is if a corridor is planned to be widened in the future, traffic signal equipment and timing could be incorporated with the construction costs. Planning for both short and long-term projects provides a region some flexibility to adjust priorities. Should there be a change in technology, travel patterns, or even policies, a region would have the ability to change projects within the plan.

The Transportation Planning Process is described in detail in *The Transportation Planning Process Key Issues A Briefing Book for Transportation Decisionmakers, Officials and Staff* published by the United States Department of Transportation (USDOT) in 2007. In short, the process is intended to create a collaborative environment that involves all users of the system and public and private agencies that design, manage, operate and maintain the transportation system. The process is illustrated in Exhibit 6 and involves the following six steps[9]:

- Monitoring existing conditions;
- Forecasting future growth and land use in the region;
- Identifying current and projected future transportation problems through planning studies and evaluating various improvement strategies to address those needs;
- Developing long-range plans and short-range programs of alternative capital improvement and operational strategies for moving people and goods;
- Estimating the impact of recommended future improvements to the transportation system on environmental features, including air quality; and
- Developing a financial plan for securing sufficient revenues to cover the costs of implementing strategies.

---

9   The Transportation Planning Process Key Issues A Briefing Book for Transportation Decisionmakers, Officials and Staff FHWA-HEP-07-039. USDOT Transportation Capacity Building Program, FHWA 2007.

Exhibit 6 Transportation Planning Process[10]

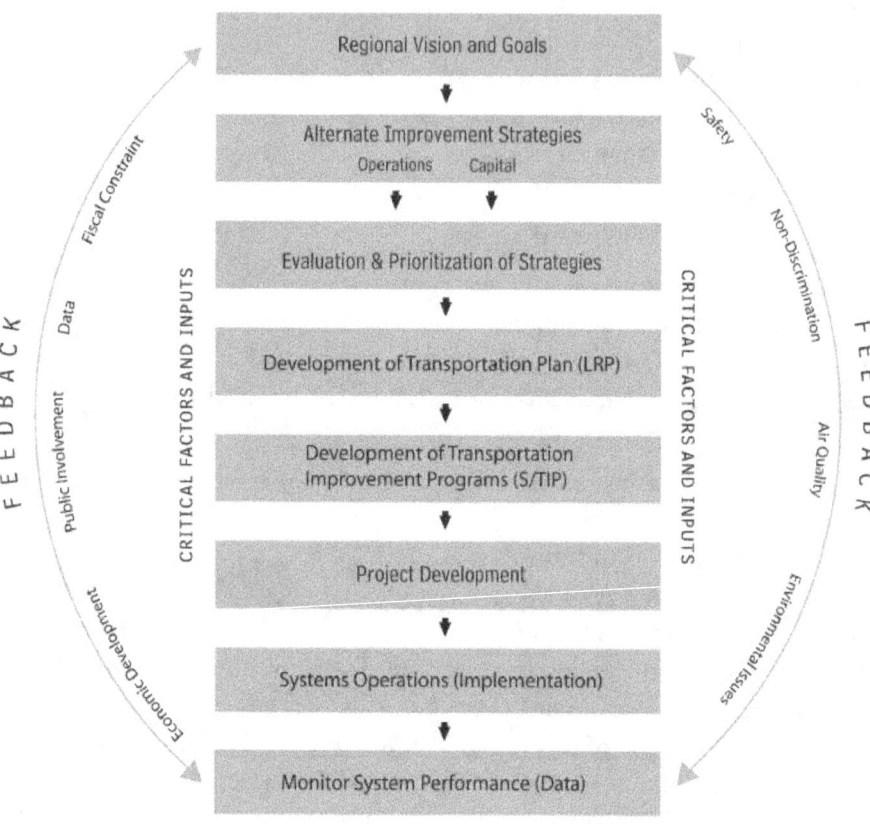

A primary role of the MPO is to seek the participation of all agencies and stakeholders within the urbanized area to build consensus on the transportation vision and goals for the region. The familiarity of MPOs with the development and monitoring of objectives and performance measures makes it uniquely qualified to facilitate Regional Traffic Signal Operations Programs. Regional operations objectives and performance measures provide a common language that allows all stakeholders in the region to agree on priorities and enables key issues related to traffic signal management to be elevated to regional significance.

Lastly, applications for regional, state, or federal funds are given greater weight when several agencies have joined together. Joint partnerships often have a higher success rate of acceptance and offer distinct cost sharing benefits. There are several examples where cooperation between neighboring agencies improves the upgrade and maintenance of projects. When funding is available, it is common for

---

10   The Transportation Planning Process Key Issues: A Briefing Book for Transportation Decision-makers, Officials, and Staff, Federal Highway Administration, Federal Transit Administration, 2007.

a regional organization or MPO to be responsible for obtaining and distributing the fund for projects that are mutually beneficial for the entire region.

Best Practices—Funding

The American Recovery and Reinvestment Act of 2009 provided an opportunity for agencies to receive federal funding to provide a stimulus to the US economy. By combining the requests across multiple agencies, the Portland-metropolitan region submitted an application for a regional arterial traffic control enhancement project to improve 277 intersections with upgrades to the signal controller hardware and software, communication, and signal timing. The project was a significant joint effort from the region to improve traffic signal management and operations. The project was an outcome of a regional TSMO plan that identified strategic investments for improving traffic signal operations.

An example of funding distribution by a regional organization is described by recent activities from the Mid-America Regional Council in Kansas City.

Best Practices—Funding

Twenty-two partner cities and agencies worked together on Operation Greenlight, a multi-year program managed by the Mid-America Regional Council to upgrade traffic signal systems throughout their metropolitan area. The major projects included signal system integration, field communication upgrades, traffic signal controller modifications and operations, and signal retiming efforts to reduce incident response time and improve air quality. The funding for the program came from more than six different sources, which indicated a commitment to the project by the represented agencies and a regional focus for the project resulting in broad involvement.

Some regions have utilized the CMAQ program to provide viable funding sources to improve signal timing and operations on a regional level. Using CMAQ funds, projects are developed to focus reducing emissions through a reduction of congestion and mobility improvements along corridors. A majority of the funds are federal dollars, but it requires a state or local match, which is commonly 20 percent.

Best Practices—Funding

The "Transportation User Fee" (TUF) used by the City of Austin, Texas, suggests another model for per-capita or per-household fees collected by the MPO. Under Austin's TUF program, municipal utility bills include a TUF, which averages $30 to $40 annually for a typical household. This charge is based on the average number of daily motor vehicle trips made per property, reflecting its size and use. The city provides exemptions to residential properties with occupants that do not own or regularly use a private motor vehicle for transportation, or if the user is 65 years of age or older.

Public Involvement and Outreach

Performance measures are an important means of developing support for the program. The performance measures must be in layman terms, considering both the needs of users and the decision makers. It is important to inform stakeholders as part of the process, especially in response to previous public feedback. This must be carefully balanced with the need to meet elected officials' expectations. Public responses and complaints can cause a reactionary response to adjust signal timing without knowing the resultant impacts. It is common to have signal timing parameters changed after receiving public complaints even though a signal timing project had been recently completed. The public's view is often narrow and taken from personal perspectives, such as the traffic signal near one's house or a corridor traveled commuting to work. Complaint responses are too commonly a result of individual response, but the impact of the individual change to the system is unknown. A few agencies and MPOs strive to change this common response by taking the time to educate the public and partaking in community outreach on the region's performance measures.

Best Practices—Public Outreach

The Denver Regional Council of Governments has an informal brochure describing the basic issues surrounding signal timing. The eight-page "Why is the Signal Always Red?[11]" is a useful document for the public that offers complaints regarding traffic signal timing.

The other aspect of public involvement is the presentation of material to the decision makers. This is critical since a majority of operational funding comes from the elected officials and their distribution of the agency budgets. A way to maintain involvement with decision makers is to engage them in regular meetings and discussions. A commitment from a decision maker can assist in a continued funding source as well as a voice in the political arena.

---

11   "Why is the signal always red?", prepared by Denver Regional Council of Governments, visited November 24, 2008, http://www.drcog.org/documents/Why%20Signal%20Red.pdf

> **Best Practices—Public Outreach**
>
> The Puget Sound Regional Council in greater Seattle, Washington, engaged the participation and involvement of a local community councilmember. The councilmember was not of traffic or a transportation background, but appreciated the importance of the many traffic signal operation related issues. The relationship between the regional council and the local councilmember allowed each party to teach and learn from each other. The councilmember learned some of the technical issues of traffic signal operations and taught how the regional group could present the information to other decision makers.

In addition to engaging elected officials as part of meetings, another way to involve decision makers is to make sure that they are well informed and aware of the regional program activities. Effective communication is a challenge as many officials are not trained with technical knowledge. It is imperative to make sure that any message communicated to any official is simple to understand and supported with facts. This is especially important as many officials are often faced with decisions or competing initiatives, such as air quality versus mobility or economic development versus quality of life concerns. Information must be provided to officials in a manner where it helps the official make a decision. Transportation operations can be related to a variety of topics that are applicable to the official's interests as well as his/her campaign promises. In addition to effectively communicating to officials, officials are dependent on staffs. Communication does not stop at the official, but includes the support staff. By communicating to the support staff, the elected official can be briefed by them when necessary.

To assist in disseminating information to various audiences, agencies employ public information officers (PIO) to effectively communicate a region's message. It is important to highlight traffic signal operations and share this information to the public through various media outlets. Outreach conducted by staff involved in signal retiming would provide an opportunity for users to understand and relate to the planned improvements. Agencies can market their efforts to change public perceptions.

Section 6
Summary

Summary

Traffic signal management and operations is a common practice for agencies across the country. Traffic signal management and operations that is collaborative across a region is not as common. The traffic signal operations community is faced with the challenge of improving the systems by working across agency boundaries with a regional perspective. It is important to recognize that neighboring agencies are faced with similar challenges, which could be met more easily by cooperating on the efforts. There are many benefits in working together, and this document provides a framework for a region to come together to collectively face those challenges.

A Regional Traffic Signal Operations Program can improve the traffic signal operations on a regional basis by bringing a diverse set of strengths together to meet the objectives of stakeholders and the communities they serve. Regional traffic signal operations produce more efficient and consistent operations that lead to improved mobility and safety. Regional programs require collaborative effort and commitment by all stakeholders. There are many benefits with collaboration as traffic signal management and operations become effective and efficient and credibility with the decision makers and the public is established.

Regional Traffic Signal Operations Programs requires partnerships among agencies working toward common objectives to be effective. Whether the objective is increasing the training of staff, improving operations across agency boundaries, or changing the paradigms to identify new funding sources, the engineers involved must interact with elected officials to introduce the program and carry it out for a region. A review of the successful programs indicates that there is no single approach to achieving regional traffic signal operations success, but there are many common themes and lessons learned that can be used for initiating a program.

www.ingramcontent.com/pod-product-compliance
Lightning Source LLC
Chambersburg PA
CBHW081850170526
45167CB00007B/2963